Fusarium Wilt of Tomato

Dr. Faramond Baier

DEDICATION

To My Wonderful Readers

CONTENTS

ACKNOWLEDGMENTS

Thank you to everyone who strives to grow and help others grow

1 ABSTRACT

The antifungal activity potentiality effects of the nanomaterial MgO and TiO2 in controlling Fusarium wilt on tomato plants under greenhouse conditions. The antifungal activity of MgO and TiO2 (NPs) were examined in vitro by using different concentration of the nanoparticles and the inhibited growth eight Fusarium isolates were recorded. Subsequently, the hyphae and spore formation were tested using the Scanning Electron Microscopy (SEM). Accordingly, greenhouse experiments were performed on three tomato cultivars inoculated by the fungus FLO4 isolate and sprayed with the desired concentrations of MgO and TiO2 compared with the control plants. The capability of the NPs in controlling the wilt disease and their prolonged effects on the plant defense response were examined. The results revealed that the two NPs succeeded to reduce the fungal radial growth and the highest reduction of the fungal growth was achieved with the high concentration of the tested NPs. A 99% reduction of fungal radial growth was obtained by MgO NPs (at 200 ppm) while 66% reduction was observed in case of TiO2NPs (at 100 ppm). The resulted in greenhouse experiments showed that MgO NPs was preferred because it decreased Fusarium wilt severity into 13.99 % compared to 51.8% infected control (F) but not treated with NPs and those treated with the chemical fungicide. Consequently, the plant immune response against the fungal infection was raised and this response was reflected through the incensement both of the activity of the defense-related enzymes (PPO and PAL) and the gene expression of PR1 and PR5 genes.

Index Terms— antifungal, defense-related enzymes, nanoparticles, fungicide, gene expression

2 INTRODUCTION

Nowadays, world is suffering from food deficiency because the tremendous increase in human population and traditional agriculture are not going to solve this problem. Moreover, drought, pests and plant diseases are additional problems which resulted in increase the gap between the produced and the demanded [1]. Plant pathogens are considered one of the most difficult problems which should be solved, and it is possible that genetic improvement of plants could be the best way to manage these losses [2]. A huge number of fungal species are acquired resistance against the most commercially available fungicides [3, 4]. Different plants infected with Fusarium wilt which is a damaging disease especially for tomato and lettuce [5].

Various natural and artificial methods for plant diseases control have been applied, among which use of fungicides is the most prevalent. Synthetic fungicides have extensively used to control phytopathogenic fungi. It well known that the use of fungicides in plant pathogen control resulted in rasing of many pesticide-resistance fungi[6] in addition disasters gained for the human health due to the residue of these fungicides accumulated in the food chain [7]. Also, fungicides caused many problems such environment contamination, elimination of non-target organisms, and selection of resistant phytopathogens [8]. Consequently, great attention is made nowadays to lessen fungicidal application in order to minimize the health hazards to man and animal. Therefore, using alternative means of disease control rather than fungicides is strongly encouraged.

A wide range of chemically and biologically synthesized nanomaterials have been used as novel antimicrobial and most of them showed resistance less for the treated plant pathogens [9, 10]. Various nanoparticles have been tested as pesticides against different plant pathogens fungi or bacteria, such as silver [11, 12], copper oxide [13, 14], titanium oxide [15, 16], zinc oxide [17], and magnesium oxide [18]. Among these plant pathogens: Botrytis cinerea, Fusariumoxysporum, Alternaria alternate, Macrophominaphaseolina, Rhizoctoniasolani, Sclerotiumrolfsii, Penicilliumspp, and Aspergillusflavus [18-21]. The aim of the present study was to determine the fungicide activity of both MgO and TiO2 nanoparticles especially against the Fusarium wilt disease and studying the plant immune response in the presence of these nanomaterials compared with the chemical synthesized pesticides.

3 MATERIALS AND METHODS

2. Materials and Methods

2.1. Isolation and identification of the causal fungi.

Eight fungal isolates were isolated from root tomato plants infected with wilt disease, which were collected from six Governorates in Egypt: El-Behiera, Alexandria, Kafr El-Sheik, El-Menofia, El-Menia and Assuit. The fungus isolates were purified by using the hyphal tip technique described by Brown [22]. The recovered fungi were transferred to PDA slants and incubated at 28°C for 7 days. Morphological identifications for the isolated fungi were carried out at the Department of Mycology and Plant Diseases, Plant Pathology Research Institute, ARC, Giza according to Gilman [23] and Summerellet al. [24]. Then the frequency of the fungal isolates recovered from the surveyed governorates was calculated.

2.2. Pathogenicity Test.

The eight fungal Fusariumoxysporum f. sp. lycopersiciisolates were tested for pathogenicity on tomato Carmen cultivar. Tomato seeds were sterilized with 1% Sodium Hypochlorite (NaOCl) for 5 min, followed by washing with dH2O, planted in 3 kg sterilized soil consists of (peat moss:clay:sand, 1:1:1) in plastic pots 20 cm diameter. Preparation of the fungi inocula were performed by culturing the pathogens on 250 mL Potato Dextrose Broth medium (PDB) 250 in Erlenmeyer flasks and the culture was incubated for 10 days at 25±2°C. After the incubation period, the culture was filtrated and the resulting conidial suspension was adjusted to 106 conidia/ml using a hemocytometer. Plants were inoculated with the pathogen conidia (50 ml) when the plant reached 15-20 cm tall according to Elad and Baker [25]. Severity of wilting was rated after 15 days of inoculation by using a scale containing 6 grades suggested by Liu et al. [26] as follows: 0= no symptoms, 1= Plants with < 25% of leaves with symptoms, 2= Plants with 25-50% of leaves with symptoms, 3= Plants with 50-75 % of leaves with symptoms, 4= Plants with 76-100% of leaves with symptoms and 5= Plants with complete death. Then, wilt disease severity was calculated as follows: Disease severity (%) = {Σ (rating no x no. plants in rating category) / (Total no. plants x highest rating value)} × 100

2.3. DNA fingerprinting of the recovered FOL isolates using Inter Simple Sequence Repeat (ISSR).

The genomic DNA for the selected fungi was isolated by wizard Genomic DNA purification kit (QIAGENE, Germany). PCR amplification was performed using six primers of Inter Simple Sequence Repeat (ISSR) to examine the polymorphism among the Fusariumoxysporum isolates [27] as shown in Table 1. PCR reaction was carried out in total volume of 25µl containing; 2.5µl 10 x buffer, 2.5 µl 50mM MgCl2, 2.5 µl 4mM dNTPs,7 µl 50pmol primer,1 µl 100 ng of fungal genomic DNA and 0.2 µl (5 units/ µl) Taq DNA polymerase (Promega Germany).The PCR reaction conditions was carried out according to Istock et al. [28]. 5 µl of the PCR product were mixed with 2µl of loading dye prior to loading of 7 µl per gel slot. Electrophoresis was performed at 100 volt with 0.5 x TBE as running buffer in 1.5% agarose/0.5x TBE gels and then gel was stained in 0.5 µg/cm3 (w/v) ethidium bromide solution and distained in deionized water. Finally the gel was visualized and photographed using gel documentation system. Presence and absence of DNA bands produced from the use of six primers were scored.

Table .1.Primers sequence used in ISSR-PCR.

Primers	Primer sequence	Annealing (°C)
UBC814	ACACACACACACACACC	55
UBC840	GAGAGAGAGAGAGAGATT	55
UBC868	GAAGAAGAAGAAGAAGAA	51
UBC835	AGAGAGAGAGAGAGAGAC	55
HB-11	GTGTGTGTGTGTCC	52
HB-13	GAGGAGGAGGC	50

2.4. Dendogram construction based on the ISSR-PCR DNA band patterns.

Data obtained by ISSR-PCR DNA band patterns were recorded for computer analysis which initially organized designated on (0-1), whereas, zero indicated absent band but 1 indicated to existed one, after excluding common bands. Pair-wise comparison of isolated fungi will be constructed (UPGMA (Un-weighted pair-Group Method with Arithmetical Averages) based on the presence or absence of unique and shared polymorphic bands and then similarity coefficients will be generated according to Jaccard [29] by Multi VariateStatisicalPakage (MVSP) Version 3.1.

2.5. MgO and TiO2 nanoparticles and their effects on Fusarium growth in vitro

The antifungal activity of the two nanoparticles (MgO and TiO2) were examined at concentrations of (50,

100 and 200 ppm) according to Elamawi and Elshafey [30]. The desired concentration was prepared by diluting the original stock solution (1000 ppm) using dH2O. The prepared solutions stored at 4oCin dark bottles until use and the sterilized H2O was used as control [31].

2.5.1. The fungicide activity of the nanoparticles on the fungal radial growth and sporulation.

The in vitro assay was performed by the agar diffusion method according to Fraternaleet al. [32]. The autoclaved PDA was mixed with the tested concentrations of nanoparticles (NPs) solutions before pouring the plates, but one was kept as control (PDA without nanoparticles). From the edge of a 7-day-old culture a 0.5 cm disc was taken of each fungus and placed in the center of each plate. Inoculated plates were then incubated at 25 ± 2oC in darkness. After the complete colonized of each fungal treatment, radial growth and sporulation of Fusarium welt fungi were determined. Four replicates were performed for each treatment. Inhibition percentages of the fungal mycelial growth (MGI) were tabulated and calculated according to Al-Taisan et al. [33] using the following formula:

$$\text{MGI (\%)} = \frac{dc - dt}{dc} \text{ x } 100$$

Where, dc and dt represent the mycelial growth diameter in control and treated plates, respectively.

2.5.2. Determination of the effect of the nanoparticles on the morphology of fungal hyphae and spores using Scanning Electron Microscopy (SEM).

Fungal hyphae and spores were scanned under the SEM (Hitachi S-3500N Scanning Electron Microscope) at the Faculty of Science, Alexandria University, and Alexandria, Egypt. To examine the effect of nanoparticles on both of hyphae and spore morphology, samples were prepared on petri dishes containing healthy10- day-old cultures on PDA medium inoculated with the tomato root rot fungi and spewed with 1 ml (100 ppm) of MgO and TiO2 nanoparticle solutions separately. On the other hand, the culture control was sprayed with 1ml sterile dH2O.

2.6. Efficacy of MgO nanoparticle in controlling of Fusariumoxysporum f. sp. lycopersici under greenhouse conditions.

The efficacy of MgO nanoparticle (200 ppm) to control of Fusariumoxysporum f. sp. lycopersici was compared with Rizolex fungicide under the greenhouse conditions. The treatments were: untreated uninfected (control), untreated infected (F), Nanoparticle treated alone (Nano), nanoparticle infected (Nano+F), Rizolex fungicide + infected (Fungicide + F). Nanoparticle solutions of MgO were prepared (200 ppm + 0.1% of sodium alginate) and when the plants were reached 15-20 cm tall (age 30 day), soil treatment was carried out by adding 50 ml of MgO (NPs) suspension to 1 kg to the root zone of seedlings of tomato before one day of the pathogen inoculation. Plants were inoculated as previously described under the pathogenicity test. Rizolex fungicide (3g/ kg) was used as a check treatment. The pots were prepared, inoculated and planted with the tomato cultivars, cv. Elisa, Carmen and super hagein, watered and fertilized as previously described under the pathogenicity test. Fungicides were applied and each treatment had 25 pots. Tomato cultivars were rated for disease severity, and Phenylalanine ammonialyase (PAL) and polyphenol oxidase (PPO) enzymes activity were evaluated in roots at 0, 0.5, 1, 3, 6, 12 and 15day post inoculation. Also, roots samples were subjected for the quantification of PR1 and PR5 genes expression using Real-Time PCR (RT-PCR) at 0, 1, 3, 6 and 15 days post inoculation, the experiment was repeated twice.

2.6.1. Effect on disease parameters.

The disease severity was assessed by the scale of [26] Liu et al. (1995) as mentioned in the pathogenicity test experiment.

2.6. 2. Effect of nanoparticles on the activity of defense related enzyme in treated tomato plants.

Three tomato cultivars (Elisa, carmen and super hagein) were used to determine the effect of nanoparticles on defense related enzyme activity of polyphenol oxidase (PPO) and peroxidase (POD). The samples were taken from around infection sit. Defense reactions were evaluated in roots after 0, 0.5, 1, 3, 6, 12 and 15 days post inoculations. Statistical analysis was made using SAS program [34].

2.6.2.1. Polyphenol oxidase (PPO) activity.

One gram of the sample was homogenized in 2 mL of 0.1 M sodium phosphate buffer (pH 6.5) in a pre-chilled Pestle and Mortar. The homogenate was centrifuged at 10,000rpm for 15 min at 4 oC and the supernatant served as enzyme source. Polyphenol oxidase activity was determined as the procedure given by Mayer et al. [35]. The reaction mixture consisted of 1.5 mL of 0.1 M sodium phosphate buffer (pH 6.5) and 200 μL of the enzyme extract. To start the reaction, 200 μL of 0.01M catechol was added. The reaction mixture was incubated at room temperature and the absorbance of the supernatant (enzyme) was set to zero at OD 495 nm using UV spectrophotometer [36]. The changes in absorbance were recorded at 30 s interval for 2 min and the enzyme activity was expressed as change in absorbance min-1 g-1 of fresh tissue [37].

2.6.2.2. Phenylalanine Ammonia-Lyase (PAL) Activity.

One gram of root sample of each replicate was homogenized in 3 ml of ice-cold sodium borate buffer (0.1 M) pH 7.0 containing 2-mercaptoethanol (1.4 mM) and insoluble polyvinylpyrrolidone (PVP) (0.1 g). The extract was filtered through cheese cloth and the filtrate was centrifuged at 15,000 g for 15 minutes at 4 °C (Universal 32R, HettichZentrifugen model D-78532, Germany). The supernatant was used as enzyme source. Phenylalanine ammonialyase activity was determined as the rate of conversion of L-phenylalanine to trans-cinnamic acid at 290 nm. About 0.4 ml of enzyme was incubated with 0.5 ml of borate buffer (0.1 M) pH 8.8 and 0.5 ml of L-phenylalanine (12 mM) for 30 min at 30 °C. The amount of trans-cinnamic acid generated was calculated using its extinction coefficient of 9630 M-1cm-1. Enzyme activity was expressed as μmol trans-cinnamic acid min-1g-1 fresh weight [38].

2.7. Quantification of the peroxidase, PR1 and PR5 genes expression using Real-time PCR in the greenhouse experiment.

2.7.1. RNA extraction and cDNA synthesis

For the gene expression analysis, roots samples were collected for the quantification of peroxidase, PR1 and PR5 genes expression using Real-time PCR (RT-PCR) at different time intervals (0, 1, 3, 6 and 15 d) post-inoculation. The samples were carried out with the experiment of the effect of nanoparticles on Fusariumoxysporium and defense-related enzyme activity in tomato cultivars in the greenhouse. The mRNA was extracted from tomato roots using BioTeke Corporation RNA Isolation kit II Guanidiumisothiocyanate Method (Maxim Biotech INC, USA) according to the manufacture procedure. The concentration and purity of RNA was determined by means of spectrophotometer and its integrity by agarose gel electrophoresis.

The first-strand cDNA was synthesizes from a reaction mixture (25µl) containing 16 µl RNase free water, 2.5 µl (5x) buffer with MgCl2, 2.5 µl (2.5 mM) dNTPs, 1 µl (10 pmol) oligo (dt) primer, 2.5 µl mRNA (2 mg/ml) and 0.5 µl Reverse Transcriptase enzyme (concentration of enzyme should mentioned and the used kit name). PCR amplification was performed in a thermal cycler programmed at 42 °C for 1 hr, 72 °C for 10 min (enzyme killing) and the product was stored at 4 °C until use.

2.7.2. Quantitative real time PCR(qRT-PCR)

Three specific primers of the expressed defense genes peroxidase, PR1 and PR5 were designed in this study obtained from Pharmacia Biotech (Amersham Pharmacia Biotech., UK Limited, HP 79NA, England) with sequence tabulated in (Table 2).The qRT-PCR reaction is performed using the Fermentase kit (Fermentas, Germany). Each reaction contained 12.5 µl of 2x Quantitech SYBR® Green RT Mix, 1µl of each forward and reverse primers (25 pm/µl), µl of the cDNA (50 ng), 9.25 µl of RNase free water for a total of 25 µl. Samples were spun before loading in the Rotor's wells. The Real-Time PCR program was a broached as follows: initial denaturation at 95 °C for 10 min.; 40 cycles of at °C for 15 sec.; annealing at 60°C for 30 sec and extension at 72 °C for 30 sec. Data acquisition performed during the extension step. This reaction was performed using Rotor-Gene- 6000-system (Qiagen, USA).

2.7.3. Real-time PCR data analysis

The relative expression of the tested gens was calculated in according to comparative CT method. ΔCq=Cq– reference gene, ΔΔCq=ΔCq – control, and ΔΔ C q expression = 2(−ΔΔCq) [39]. The expression levels of the target genes were normalized relative to 18S rRNA gene and relative expression of untreated control plants at each time were set as 1.

Table 2: The primer sequence used in the Real-Time was as follows:

Primers	Primer sequence	Annealing (°C)
PR1	F: 5'-TTCAAGGCTACTCTGGC-3'	60
	R: 3'CAAGCCATTGTGGAGAT 5'	
PR5	F: 5'CATGCTCTTGATGAGGCGTA 3'	60
	R: 3'CCATCTATGGAACGGGAAGA5'	
18S	F: 5'-GTGCATGGCCGTTCTTAGTTG-3'	60
	R: 5'-CAGGCTGAGGTCTCGTTCGT-3'	

2.8. Statistical analysis

Data were statistically analyzed by two-way analysis of variance (treatments and times) as repeated measures over time using the MIXED procedure of the statistical analysis software (SAS) version 9.4 Cary, NC, SAS Institute Inc. [34] . Least significant means were compared using Dunnett's post-hoc Test ($P < 0.05$).

3. Results

Isolation and identification of the associated tomato Fusarium wilt. Eight Fusariumoxysporum f. sp.lycopersici isolates of tomato Fusarium wilt were isolated from infected tomato collected from El-Menofia (Talla (FOL1) and shebeenElkom (FOL2)) , El-Menia (FOL3), Assuit (FOL4), Kafr El-Sheikh (FOL5), and Alexandria (FOL6), El-Beheira governorate (abou-Homus (FOL7) and Kom Hamada (FOL8).

3.2. Pathogenicity test

Pathogenicity test of the Fusariumoxysporum f. sp. Lycopersici isolates were performed on Carmen tomato cultivar. Data presented in Table (2) showed that all isolates were virulent; isolates of FOL4, FOL2 and FOL5 were highly virulent produced 75%, 70% and 66% respectively, isolates of FOL6 and FOL8 were moderately virulent produced 50% and 46% respectively and the isolates of FOL7, FOL3 and FOL1 were weakly virulent produced 30%, 25% and 13% respectively.

Fusariumisolates	Diseased severity %
FOL4	75a
FOL2	70a
FOL5	66a
FOL6	50b
FOL8	46b
FOL7	30c
FOL3	25c
FOL1	13d

Table 3: Pathogenicity on tomato cv. Carmen of eight of Fusariumoxysporum f. sp. lycopersiciisolates

recovered from tomato plants showed Fusarium wilt from six major governorates for tomato cultivation during the 2019 growing season.

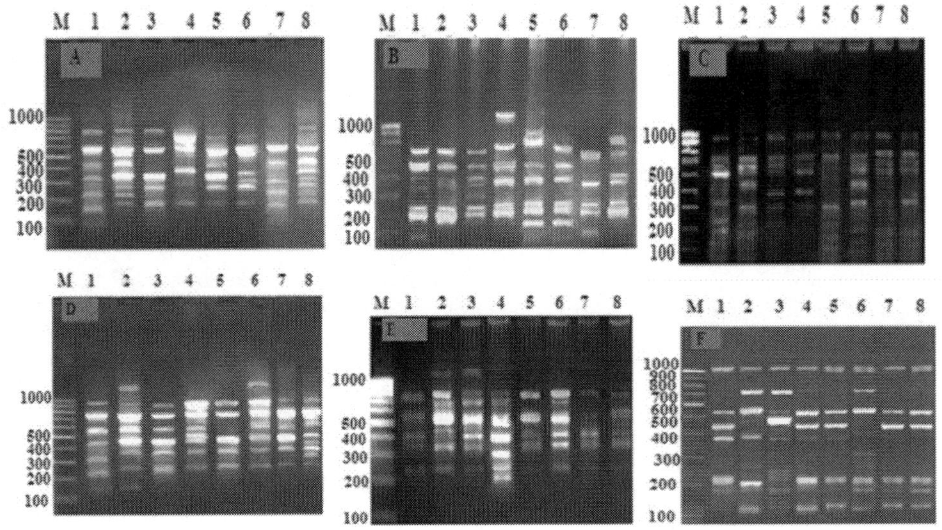

* Data are average of three replicates.

3. 3. Fingerprinting of isolates using Inter Simple Sequence Repeat (ISSR)

Using six primers in ISSR showed different level of polymorphism among the eight studied Fusariumoxysporumisolates on the basis of amplified product band patterns observed with each primer. Also, high similarity was observed between FOL3 and FOL4. The amplification profiles with the primers UBC814,UBC840, UBC868, UBC835, HB-11and HB-13 are showed in Figure(1) revealed that all of these primers succeed to give polymorphic patterns among the isolates.

Fig.(1):ISSR-PCR using primer A: UBC814, B: UBC840, C: UBC868, D: UBC835, E: HB-11 and F: HB-13. M: DNA marker. Lane 1:FOL1, Lane 2:FOL7, Lane 3: FOL3, Lane 4: FOL5, Lane 5:FOL6, Lane 6:FOL2, Lane 7:FOL8 and Lane 8:FOL4

3.3.1. Cluster analysis of ISSR results

The ISSR band patterns were analysed using UPGMA method to construct a similarity matrix to generate a dendrogram indicating the relationship between the eight examined Fusarium isolates. The presence or absence of any particular DNA bands was the only factor considered in the computer analysis. The dendrogram generated show linkage distance between the isolates Figure (2) indicated that the examined isolates were classified into Three main Clusters, (Cluster A, Cluster B and Cluster C). Cluster A split into two sub- clusters, sub-cluster (A1) divided into two groups: group (1) split into two subgroups: subgroup (1) include FOL7 isolate, subgroup (2) includes FOL4, FOL3 and FOL5 isolates and group (2) contain FOL8 isolate. Sub-cluster (A2) includes FOL6 isolate, however cluster (B) include FOL1 isolate and cluster C include FOL2 isolate.

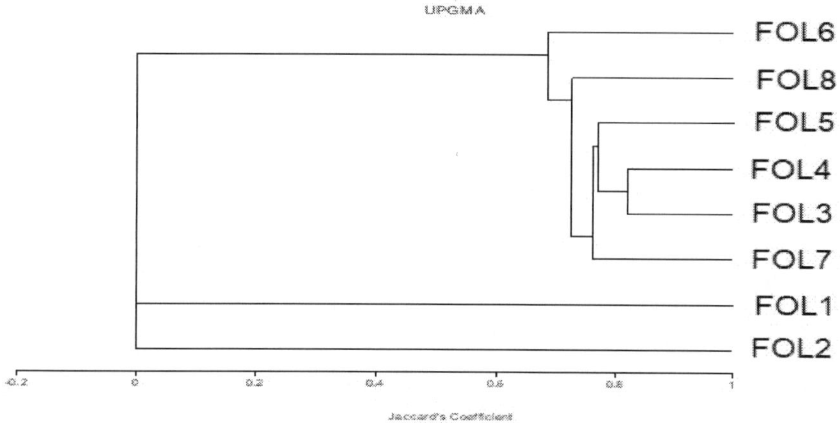

Figure(2):Dendrogram show phylogenetic relationship among the eight Fusariumoxysporium based on ISSRs analysis.

3.4. The in vitro effect of the two nanoparticles on the tomato Fusariumwilt fungi

3.4.1. Effect on fungal radial growth

Data presented in Table (4) and figure (3) showed that all the different concentrations from MgO and TiO2 nanoparticles (50, 100 and 200 ppm) caused significantly inhibition on the fungal radial growth of the tested fusarium isolates on PDA with different degrees compared to the control. Also, data showed that the high inhibition activity for all tested fungi obtained with concentration of (200 ppm) of Mgo nanoparticles was 82.28%. While the maximum inhibition activity of TiO2 nanoparticles at (100 ppm) concentration was 64.69 % (Table 3). So the MgONps showed considerable antifungal activity against Fusarium isolates than in TiO2 Nps.

Table (4): The in vitro percentage of inhibition of radial growth of tomato Fusarium wilt pathogenic fungi on PDA amended with nanoparticles MgO and TiO2 at different concentrations.

Nanoparticles / Isolate	MgO (ppm)			TiO2 (ppm)		
	50	100	200	50	100	200
FOL4	28.75	48.75	76.25	27.5	61.25	47.5
FOL2	37.5	50	78.75	30.875	62.5	47.5
FOL5	43.75	56.25	80	31.25	62.5	56.25
FOL6	46.25	60	81.25	32.5	62.5	57.5
FOL8	46.25	60	82.5	37.5	62.5	60
FOL7	47.5	60	83.25	43.75	65	60
FOL3	50	62.5	86.25	47.5	66.25	61.25
FOL1	51.25	67.5	90	53.75	75	66.25
Mean	43.91c	58.13b	82.28a	38.1c	64.69a	57.03b
	LSD0.05 (Treat.) = 7.36			LSD0.05 (Treat.) = 5.18		

* Inhibition effects were determined based on four replicates for each treatment. Values followed by different letter are significantly different at p=0.05

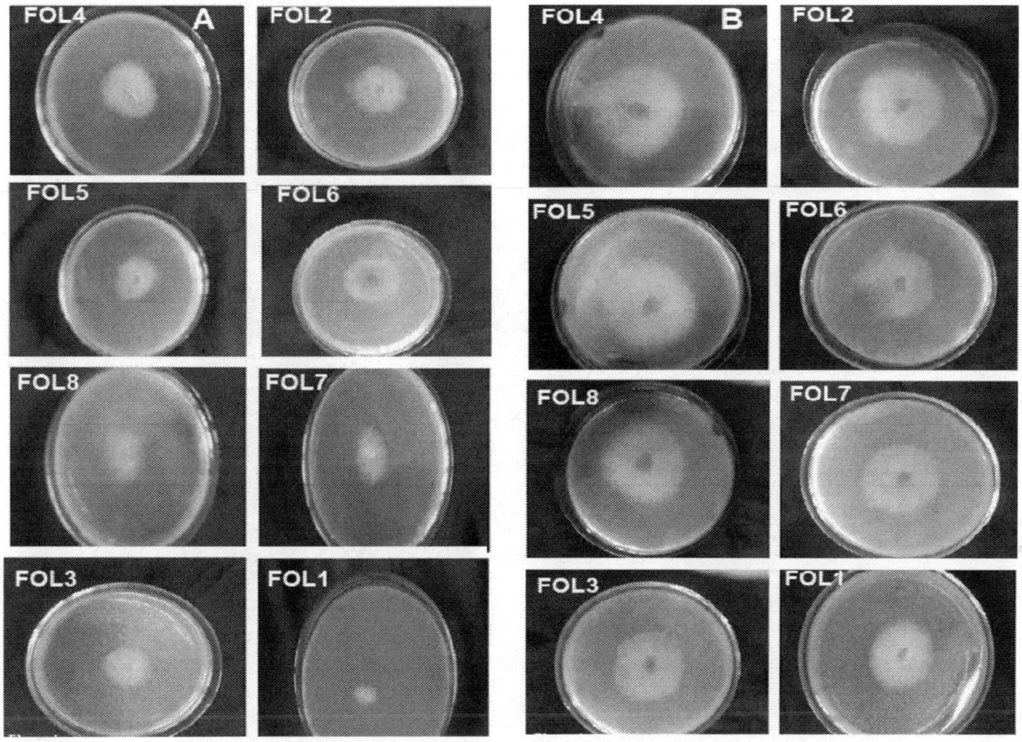

Fig.(3):The Effect of MgO nanoparticles at 200ppm concentration (A) and TiO2 nanoparticles at 100ppm concentration (B) on inhibition of radial growth of Fusariumoxysporum f. sp. Lycopersicion PDA.

3.4.2. Effect of nanoparticles on the hyphal and spore morphology of FOL isolates.

In order to reveal the effect of the tested nanoparticles on hyphal and spore morphology, healthy fungal 10 days old cultures on PDA of the tomato Fusarium wilt fungi were sprayed with 100 ppm MgO, 200 ppm TiO2 and other plates were also, sprayed with deionized water as control and investigated by Scanning Electron Microscope (SEM) 24 hours after treatment. It was evident from Fig. (4) that in the control, the

SEM investigation showed the images of conidia and mycelia of Fusariumoxysporum f. sp. Lycopersiciwith typical net structure and smooth surface while in the treated cultures with MgO and TiO2nanoparticles (NPs) clearly damaged the hyphae and conidia of Fusariumoxysporum f. sp. Lycopersicias fungal mycelia and conidia were sunken, wrinkled and damaged (Fig. 11). In contrast MgO NPs formed unusual bulges on the surface of fungal hyphae while, TiO2 caused severe hyphal distortion.

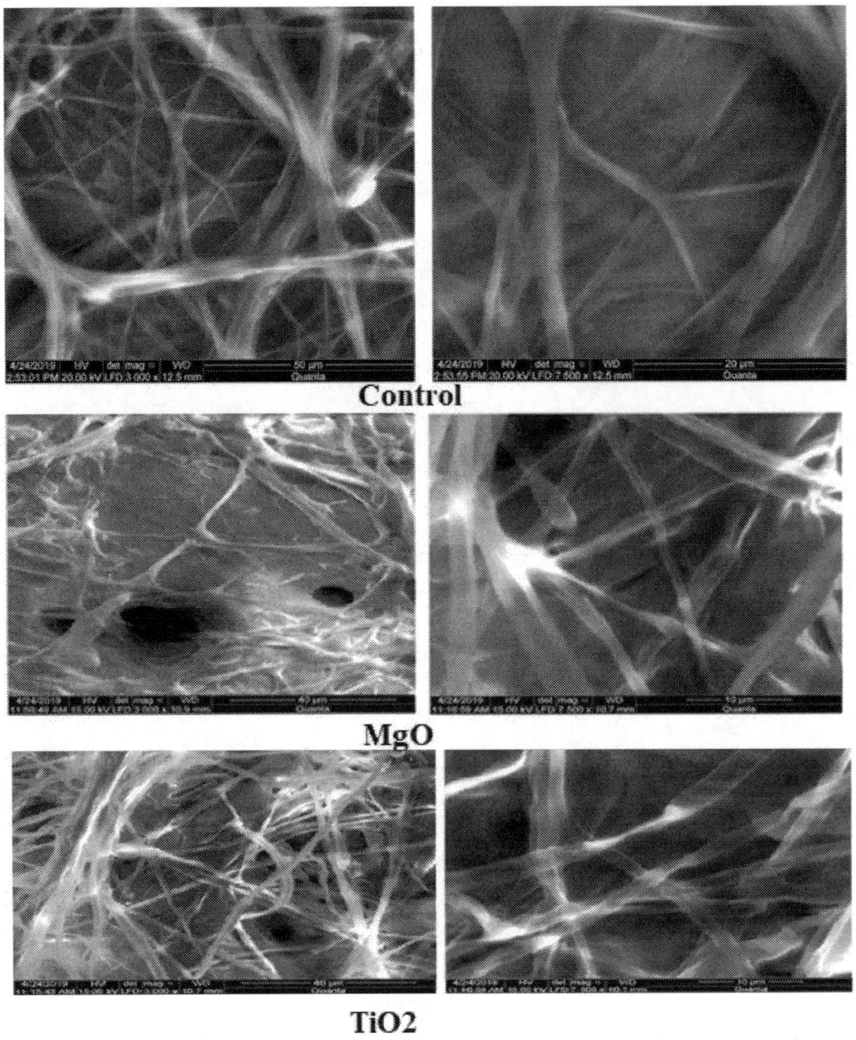

Fig.(4): Scanning Electron Microscopy (SEM) of conidia and hyphae of Fusariumoxysporum f. sp. Lycopersici (FOL1) grown on PDA sprayed with water as a control or 200 ppm of MgO, and 100 ppm of

TiO_2 Nanoparticle solutions, 24 h after treatment.

3.5. Efficacy of MgO nanoparticle to control of Fusariumoxysporum f. sp. lycopersiciunder greenhouse.

3.5.1. Effect on disease severity

The results explain that the three tomato cultivars used in the study was sensitive to Fusarium pathogen by achieving disease incidence (51.87%). Also the three tomato cultivars exhibited different capabilities in reducing the disease severity percentage at 200 ppm of MgO nanoparticles (NPs) concentration. According to data presented in Table (4) The 200 ppm MgONps showed the high inhibition for all tested fungi was (82.28%) and rizolex fungicide on Fusarium wilt. The disease severity on tomato cultivars; Elisa, Carmen and super hagein caused by Fusariumoxysporum f. sp. Lycopersiciis shown in Table (5) and fig (5). Generally MgO nanoparticles (Nano+F) and rizolex(Fungicide+F)significantly decreased fusarium welt severity. Treatment with Nano+F was the most effective and decreased mean percentage of fusarium welt severity to 13.95 % compared to 51.79% for the untreated inoculated control (F). This was not significantly different from the Nano+F effect and Fungicide+F. The TiO_2 effect was not significantly different from the Vitavax fungicide effect. For the post-emergence damping-off, MgO was most effective and reduced post-emergence to 5.3% compared to 22.1% for the untreated control. This effect was not significantly different from TiO_2 and the fungicide Vitavax.

Table (5): The effect of (200ppm) MgONps and fungicide rizolex on percentage of tomato Fusarium wilt disease severity.

	Treatments				
cultivars	F	rizolex+F	MgO(Nano)+F	Nano	Control
Elisa	33.70	10.96	11.64	0	0
Carmen	51.68	11.10	14.08	0	0
Super Hagein	70.24	14.24	16.26	0	0
Mean	51.87a	12.10b	13.99b	0	0

*Fusarium welt severity was determined based on three replicates for each treatment. LSD0.05 = 6.74

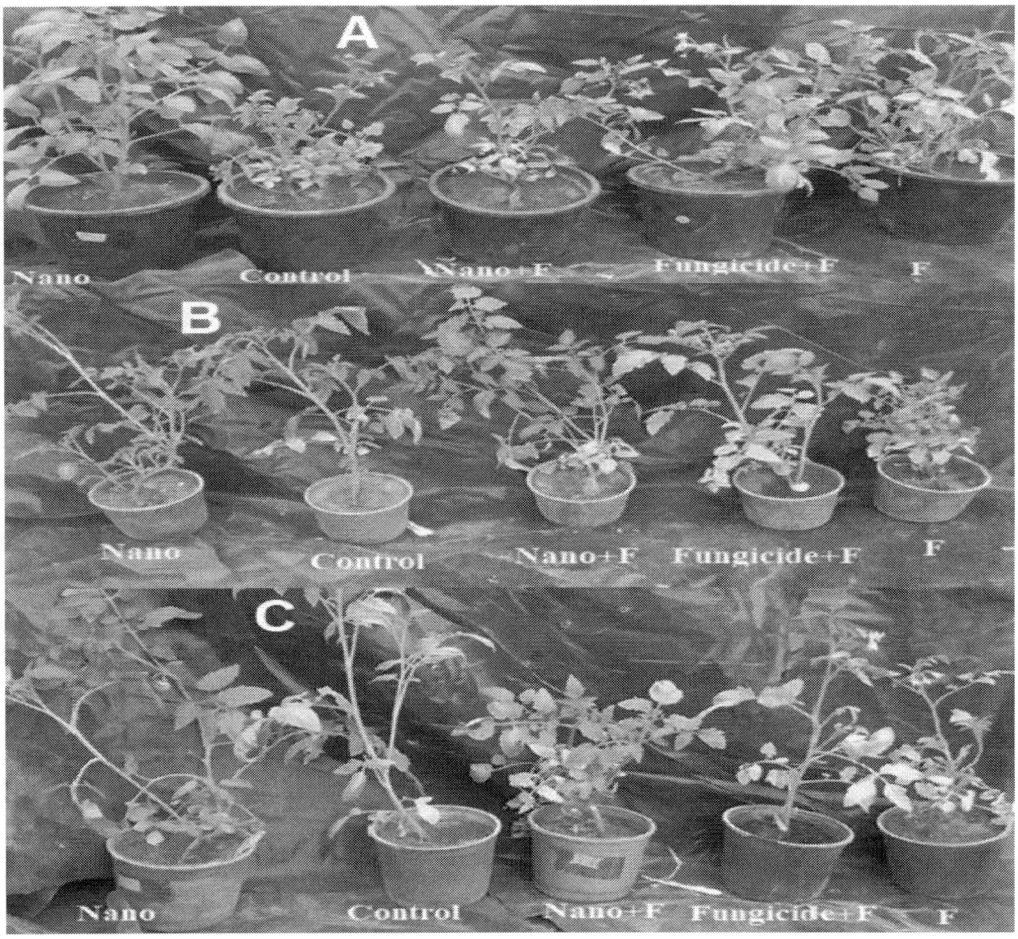

Fig. 5: The effect of MgO or TiO2 nanoparticles on fusarium wilt disease severity on Tomatoes cultivars infected by tomato fusarium wilt (A: Elisa, B: Carmen and C: Super hagein)

3.5.2. Effect of nanoparticles on defense related enzymes activity.

Three tomato cultivars; Elisa and super hagein were used to test the defense related enzyme activity of peroxidase (POD), and polyphenol oxidase PAL. Defense reactions were evaluated in tomato roots 0, 0.5, 1, 3, 6, 15 d post inoculations by FOL4 fusarium isolate after the treatment with 200 ppm Mgo nanoparticle concentrate+fusarium (Nano+F), Nano, rozalex (Fungicide+ F), F and control.

Data presented in Fig. (6) revealed that PPO and PAL activities in tomato roots of Elisa, carmen and super hageincvs increased with all treatments compared with the control. As a mean of treatments, Nano+F and Fungcide+F were the most effective followed by F, while no significant between nano and control on increasing the PAL and PPO activity. Results also showed that, the highest increase of PPO activity was observed at 3 d in tomato roots of Elisa, carmen and super hageincvs, then decreased after 6 to 15 d , while the highest increase of PAL activity was observed at 6 d in Elisa, carmen, while at 15 d in super hagein. Data showed that the highest increase of PPO activity was observed in Elisa c.v followed by carmenc.v, while the low activity was observed in super hagein c.v. These results are in harmony with El-Argawyet al., [31] as they found that treatments with the testedTiO2, MgO and ZnO NPs and Vitavax enhanced the plant defense and related to the enzyme activity i.e., polyphenol oxidase and the total phenols compared to the untreated infected control.

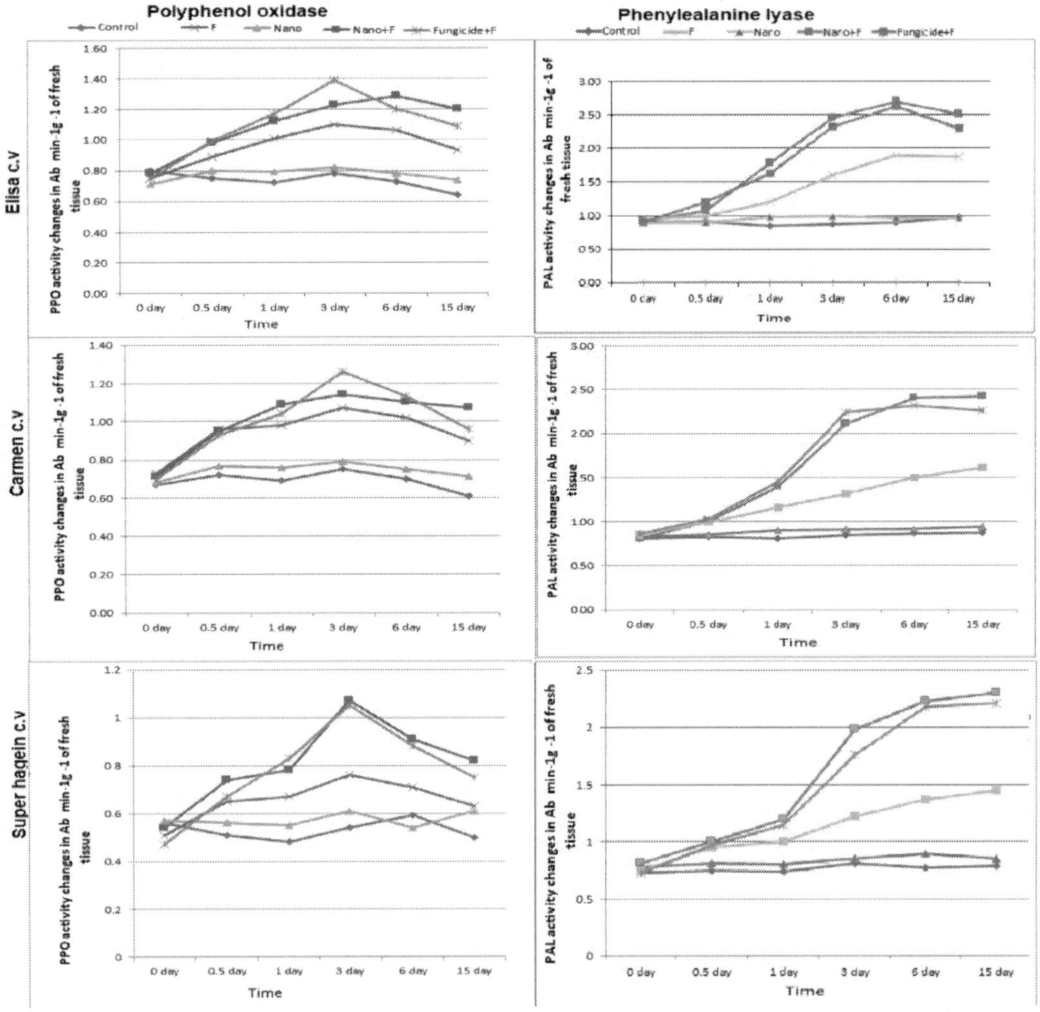

Fig. 6: PPO and PAL activity in tomato roots of Elisa, Carmen and Super hageinc.vs after treatment with Mgo NPs followed by inoculation with Fusariumoxysporum f. sp. Lycopersici.

3.6.Quantification of the Peroxidase POD, PR1 and PR5 genes expression using Real-Time PCR(qRT-PCR)

Data presented in Figs (7) revealed that the level expression of PR1 and PR5 gene in tomato roots of; Elisa, Carmen and Super hageincvs was increased by treatments with (200ppm MgONps) compared to the control. The level of expression of The PR1 gene for Elisa and Carmen in Nano+F was highly expressed followed by Fungicide+F, F and Nano compared the control, while the treatment Fungicide+F was the most effective with super hagein followed by Nano+F, F and Nano compared the control. The high PR5 gene expression was observed in Nano+F was the most effective followed by Fungicide+F , F and Nano compared the control for all cultivars.

Data also indicated that the highest increase in expression of PR1 gene was reached the maximum level at 15 d with Elisa and Carmen, while the maximum level of POD gene expression at 6 d in super hageincvs in all treatments. But the highest increase in expression of PR1 gene with (Fungicide+F) was reached the maximum level at 3 d in Carmen cvs. Data also indicated that the highest increase in expression of PR5 gene was reached the maximum level at 15 d in Elisa and 6d in Carmen and Super hageincvs. The highest increase in level of expression of PR1 and PR5 genes were observed in tomato roots of Elisa c.v compared to Carmen and Super hageincvs.. Also, Gene expression in watermelon roots revealed strong upregulation of polyphenol oxidase (PPO) and PR1 genes when CuONP and F. oxysporumf. sp. niveum were both present. Enzymatic assays for PPO supported the gene expression results. CuO NP may serve as a highly effective delivery agent for this micronutrient to suppress disease.

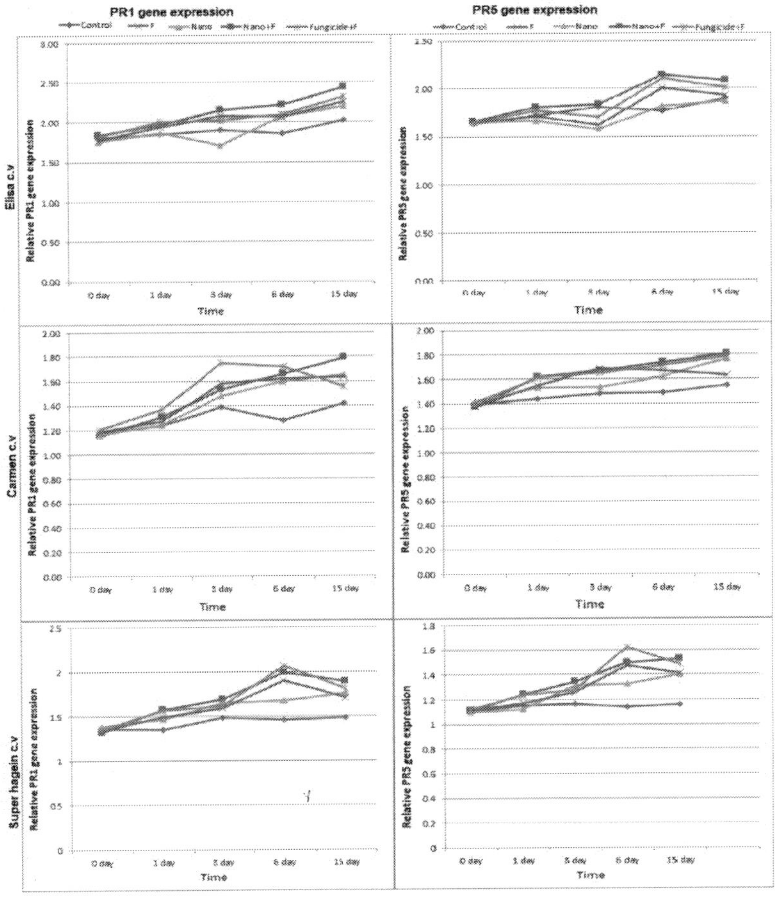

Fig. 7: Relative PR1 and PR5 genes expression in Elisa, Carmen and Super hageinc.vs tomato roots after treatment with MgO NPs followed by inoculation with Fusariumoxysporum f. sp. Lycopersici.

4. Discussion

Tomato is one of the leading and economic vegetable crops in the world, about 130 million tons all the amount of tomato yield/year [40].Tomato are infected with many diseases among of them is the wilt disease which caused by Fusariumoxysporum f. sp. lycopersici. This disease is highly destructive vascular wilt disease and caused a high loss in the crop yield [41](McGovern, 2015).Isolation of 8 isolates of Eight Fusariumoxysporum f. sp. lycopersici from six different governorates from tomato filed indicated that this disease is highly distributed in Egypt especially in Delta and Upper Egypt. The same observation was reported by [42]Abou-Zeid et al. (2016), they isolated six differently pathogenic isolates of Fusariumoxysporum from Dakahlia and Gharbia governorates, in Egypt. In addition, they found that the isolate F3 of F. oxysporum collected from Dakahlia samples was more aggressive than Tanta isolate F4. Here in this study three isolates showed high pathogenicity (aggressive) and three were moderate and three are weak. The variability of pathogenicity among the 8 examined fungi didn`t reflect the genetic polymorphism obtained by ISSR-PCR. It was observed that isolates 8, 5, 4, 37 are grouped in one group while isolate 6, 1 and 2 each one of them form one group in separate manner. In contrast differences in ISSR polymorphism among the Fusarium isolates were not in accordance with the fungus pathogenicity isolated from either tomato or cabbage [43].

For avoiding the crop losses caused by different plant diseases and for controlling the resistance of phytopathogenic fungi to chemical fungicides in addition to the hazardous gained by these fungicides [44](Ahmed and Lee, (2015). So, scientists are in contentious work searching on new materials which could be able to control most of the plant diseases especially fungi and break their power full of fungicides resistance [45] (Bouwmeester et al., 2009) through development of nanopesticides. The same idea was approached by Sekhon, (2014)[46] when used the nanoparticles (NP) to control phytopathogens. Here it was observed that the using of MgO and TiO2 nanoparticles in controlling of Fusariumoxysporum f. sp. lycopersici gave high fungicide activity compared with the chemical fungicide. Moreover, the fungicidal activity of the MgONps was higher than that obtained by TiO2 Nps when both were applied on the examined 8 fungal isolates. This finding are in agree with Abdel-Aziz et al. [47]who reported that MgO nanoparticles have high impact in controlling of Fusariumoxysporum F. Sp. lycopersici. El-Abeid et al. [48] study using the MgO in plant protection against plant wilt disease. TiO2 was used as antifungal materials by Servin et al. [49] 2015) and they observed that these particles induced the plant immune system against the plant pathogen. Liu et al. [50] revealed that TiO2 have antifungal activity against fungi Fusariumgraminearum. Both MgO and TiO2 nanoparticles inhibited the Fusariumoxysporum f. sp. Lycopersici growth in addition, affected the hyphae and candida characters. Moreover, male formation in the spore morphology was observed.

Our findings were in agreement with Ahmed et al. [51] who reported that Nickel and Magnesium Oxide inhibited the fungal radial growth of F. oxysporum f. sp. lycopersiciand El-Argawy et al [31] as reported that MgO, TiO2 and ZnO nanoparticles (NPs) inhibited the fungal radial growth of F. oxysporum f. sp. betae. Meanwhile, TiO2 NP (100 ppm) showed the highest effect in decreasing mean radial growth. Also, TiO2 showed 100% inhibition on sclerotia formation of Sclerotiumrolfsii, while MgO was most effective and decreased sporulation (number of conidia) of Fusariunoxysporum f. sp., betae. El-Argawy et al. [31] reported that the SEM investigation showed the treated cultures MgO and TiO2 nanoparticles (NPs) clearly damaged the hyphae and conidia of F. oxysporum f. sp. betae as fungal mycelia and conidia were sunken, wrinkled and damaged. In contrast MgO NPs formed unusual bulges on the surface of fungal hyphae while, TiO2 caused severe hyphal distortion.El-Argawy et al. [31] who reported that the treatment with TiO2 Np was the most effective and decreased percentage of damping-off to 16.60% compared to 38.8% for the untreated control.Elamawi and Shafey [30] reported that Fusarium disease incidence on tomato was reduced by silver nanoparticles to 5% compared to 100% for the untreated control. These results are in harmony with El-Argawy et al., [31] as they found that treatments with the tested TiO2, MgO and ZnO NPs and Vitavax enhanced the plant defense and related to the enzyme activity i.e., polyphenol oxidase and the total phenols compared to the untreated infected control. El-Abeid et al. [48] reported that MgO induce the plant immune response through the jasmonic acid signaling pathway.

On the other hand, Imada et al. (2016) [52] postulated that MgO nanoparticles are capable to induce the plant systemic resistance against the bacterial wilt disease in tomato. Additionally, it was reported that MgO NPs can increase fresh weight of wheat and carotenoid concentration as well [53]. They added that MgO could be used as fertilizer in agriculture practice. Meanwhile, Gohari et al. (2020)[54] reported that TiO2 can promot the plant growth and improve the plant resistance against salinity stress. Due to the interfering of the MgO and TiO2 in plant response against the plant pathogen and due to their effects on the plant defense system, it was observed that the plant immune system was induced by the nanoparticles of the two elements. They trigger this type of defense through the path ways of both jasmonic acid and salicylic acid.

It was observed in this study and in the greenhouse experiment that both of MgO and TiO2 nanoparticles when applied on the tomato plants infected with Fusarium wilt disease, the disease severity was decreased in compared with the used chemical fungicides. Moreover, the MgO was the most effective fungicide among the other tested fungicides examined in this study. We assumed that both the examined nanoparticles succeed in inhibition the pathogenic fungi and induced the plant defense against the pathogen in the same time, which reflected on the enzymes high activity and enhanced the plant physiological parameters; also the increasing of the expression of plant defense gene such as PR family. The result indicates that expression of these defenes genes was enhanced with MgONps which induce the plant defense system against Fusarium infection. These results are in harmony with Elmer et al., [55] as the foliar sprays with nanoparticles (NP) of CuO,MnO, SiO, TiO, and ZnO to protect watermelon against Fusarium wilt. This assumption are in agree with Siddaiah et al. [56] who reported that the seed treatment with Chitosan nanoparticles showed changes in defense gene expression profiles where in expression of genes of phenylalanine ammonia lyase (PAL), peroxidase, polyphenol oxidase (PPO), catalase and superoxide dismutase were highly upregulated. Chitosan nanoparticles treatment resulted in earlier and higher expression of the pathogenesis related proteins PR1 and PR5.

5. Conclusions

Although tomatos wilt diseases can be controlled by synthetic fungicides. But the use of fungicides has been restricted due to their carcinogenicity, high and acute residual toxicity, long degradation period, environmental pollution and side-effects on human health. For this reason, new alternatives have been explored to reduce the use of synthetic fungicides. The current study showed that the treatment of tomato Fusarium wilt by Magnesium oxide NPs decreased the percentage of Fusarium wilt severity and increases the activity and gene expression of the defense-related enzymes. The use of nanoparticles as an environmentally safe substitute for the traditional synthetic fungicide will provide a green method of controlling plant diseases. More attention should be given to developing a strong delivery method of nanoparticle compounds in different agriculture systems. For these results we can use new alternatives to reduce the use of the fungicides.

Institutional Review Board Statement: Not applicable.
Informed Consent Statement: Not applicable.
Data Availability Statement: Relevant data applicable to this research are within the paper.
Conflicts of Interest: All authors declare no conflict of interest.

References

[1] Bowyer, P. 1999. Plant disease caused by fungi: phytopathogenicity In: Molecular Fungal Biology, (RP Oliver, M Schweizer), Cambridge University Press, Cambridg. https://www.research.manchester.ac.uk/portal/en/publications/.html

[2] Ashkani, S, M. Rafii, M.Shabanimofrad, A. Ghasemzadeh,S. Ravanfar and M. Latif. 2014. Molecular progress on the mapping and cloning of functional genes for blast disease in rice (Oryza sativa L.): current status and future considerations.Crit. Rev. Biotechnol., 8: 1–15. https://pubmed.ncbi.nlm.nih.gov/25394538/

[3] Goffeau, A. 2008. Drug resistance: The fight against fungi. Nature, 452: 541-542.

[4] https://www.nature.com/articles/452541a

[5] Omar, A. M and Ahmed I.S. A. (2014). Antagonistic and inhibitory effect of some plant rhizo-bacteria against different Fusarium isolates on Salvia officinalis. American-Eurasian J. Agric. & Environ. Sci., 14 (12): 1437-1446

[6] Srinivas, C, D. N. Devi, K. N. Murthy, C. D. Mohan, T.R. Lakshmeesha , B.P. Singh, N.K. Kalagatur, S.R. Niranjana, A. H. Abdulaziz, A. A. lqarawi, B. Tabassum, E. F. Abd_Allah, S. Ch. Nayaka and R. K. Srivastava. 2019. Fusariumoxysporum f. sp. lycopersici causal agent of vascular wilt disease of tomato: Biology to diversity– A review. Saudi Journal of Biological Sciences 26:1315–1324. https://www.sciencedirect.com/science/article/pii/S1319562X19301007

[7] Staub, T. 1991. Fungicide resistance: practical experience with antiresistance strategies and the role of integrated use. Annu. Rev. Phytopathol., 29:421-442. https://pubmed.ncbi.nlm.nih.gov/18479197/

[8] El-Nahhal, Y. 2004.Contamination and safety status of plant food in Arab countries. J. Appl. Sci. 4: 411-417. https://scialert.net/fulltext/?doi=jas.2004.411.417

[9] Mpumi, N, R. S. Machunda, K. M. Mtei and P. A. Ndakidemi .2020. Selected Insect Pests of Economic Importance to Brassica oleracea, Their Control Strategies and the Potential Threat to Environmental Pollution in Africa. Sustainability, 12, 3824. Downloads/sustainability-12-03824-v2.pdf

[10] Morones, J. R., J. L. Elechiguerra, A. Camacho, K. Holt, J. B. Kouri, J. T. Ramirez and M. J. Yacaman. 2005.The bactericidal effect of silver nanoparticles. Nanobiotechnology, 16:2346-2361. https://pubmed.ncbi.nlm.nih.gov/20818017/

[11] Phogat N., A.S. Khan., S. S., Ansary, A. A., Uddin. 2016. Fate of inorganic nanoparticles in agriculture. Adv. Mater. Lett. 7, 03–12. https://www.researchgate.net/publication/291354035_Fate_Of_Inorganic_Nanoparticles_In_Agriculture

[12] Kumar,V. and S. K. Yadav. 2009. Plant- mediated synthesis of silver and gold nanoparticles and their application. J. of Chemical Technology and Biotechnology, 84: 151-157. https://onlinelibrary.wiley.com/doi/full/10.1002/jctb.2023

[13] Panácˇek A, M.Kolár, R. Vecerová, R. Prucek, J. Soukupová. and V. Krystof. 2009. Antifungal activity of silver nanoparticles against Candida spp. Biomaterials 30, 6333–6340. https://www.sciencedirect.com/science/article/abs/pii/S0142961209008023

[14] Cioffi, N., L. Torsi, N. Ditaranto, G. Tantillo, L. Ghibelli and L. Sabbatini. 2005. Copper nanoparticle/polymer composites with antifungal and bacteriostatic properties. Chem. Mater. 17: 5255-62. https://pubs.acs.org/doi/10.1021/cm0505244

[15] Kanhed.P, S. Birla, S. Gaikwad, A. Gade, A. B. Seabra, O. Rubilar.2014. In vitro antifungal efficacy of copper nanoparticles against selected crop pathogenic fungi Materials Letters, 115 : 13-17. https://www.sciencedirect.com/science/article/abs/pii/S0167577X13013888

[16] Kwak, S.Y., S. H. Kim and S. S. Kim. 2001. Hybrid organic/inorganic reverse osmosis (RO) membrane for bactericidal antifouling. 1. Preparation and characterization of TiO2 nanoparticle self-assembled aromatic polyamide thinfilm- composite (TFC) membrane. Environ. Sci. Technol. 35:2388-94. https://pubmed.ncbi.nlm.nih.gov/11414050/

[17] Gutierrez, F. M., P. L. Olive, A. Banuelos, E. Orrantia, N. Nino, E. M. Sanchez, F. Ruiz, H. Bach, and Y. A Gay. 2010. Synthesis, characterization, and evaluation of antimicrobial and cytotoxic effect of silver and titanium nanoparticles. Nanomed. 6:681-688. https://www.sciencedirect.com/science/article/abs/pii/S154996341000095X

[18] Liu, Y., L. He, A. Mustapha, H. Li and M. Lin. 2009. Antibacterial activities of zinc oxide nanoparticles against Escherichia coli O157: H7. J. Appl. Microbiol. 107: 1193-201. https://sfamjournals.onlinelibrary.wiley.com/doi/pdf/10.1111/j.1365-2672.2009.04303.x

[19] Wani, A. H., M. Amin and M. Shahnaz. 2012. Antimycotic activity of nanoparticles of Mgo, FeO and ZnO on some pathogenic fungi. Inter. J. Manufact. Mater. Mechanic. Engineer. 2: 59-70. https://ideas.repec.org/a/igg/jmmme0/v2y2012i4p59-70.html

[20] He, L., Y. Liu, A. Mustapha, and M. Lin. 2011. Antifungal activity of zinc oxide nanoparticles against Botrytis cinerea and Penicilliumexpansum. Microbiological Research, 166: 207-215. https://europepmc.org/article/med/20630731

[21] Lamsal, K., S. W. Kim, J. H. Jung, Y. S. Kim, K. S. Kim and Y. S. Lee. 2011. Inhibition effects of silver nanoparticles against powdery mildews on cucumber and pumpkin. Mycobiol. 39: 26-32. https://www.ncbi.nlm.nih.gov/pmc/articles/PMC3385079/

[22] Yehia, R. S. and O. F. Ahmed. 2013. In vitro study of the antifungal efficacy of zinc oxide nanoparticles against Fusariumoxysporum and Peniciliumexpansum. Afr. J. Microbiol. Res. 7: 1917-1923. https://academicjournals.org/journal/AJMR/article-abstract/3F15F4B11762

[23] Brown, N. 1924. Two mycological Method .II a Method of Isolated Single Strain Fungi by cutting a hyphal tip. Ann .Bot. 38:402-406. https://academic.oup.com/aob/article

[24] https://doi.org/10.3389/fmicb.2019.00616

[25] Gilman, J. C. 1957. A manual of soil fungi Iowa State University Press. Ames. Iowa. USA,pp450.https://journals.lww.com/soilsci/citation/1957/08000/a_manual_of_soil_fungi.21.aspx

[26] Summerell, B. A., B. Salleh and J. F. Leslie. 2003. Autilitrarian approach to Fusarium identification. Plant Dis. 87: 117-128. https://pubmed.ncbi.nlm.nih.gov/18479197

[27] Elad, Y. and R. Baker. 1985. The role of competition for iron and carbon in suppression of chlamydospore germination of Fusarium spp. by Pseudomonas spp. J. Phytopathol. 75: 1053-1059. https://www.researchgate.net/publication/249298514_

[28] Liu.L, J.W. Kloepper, S. Tuzun. 1995. Induction of systemic resistance in Cucumber against Fusarium wilt by plant growth-promoting rhizobacteri, Phytopathology 85 :695-698. https://europepmc.org/article/agr/ind20527968

[29] Tabikha, R.M., Adss, I.A. Morphologic and genetic analysis for geographic populations of greenbugSchizaphisgraminum (Hemiptera: Aphididae) in Egypt. Biologia 76, 77–89 (2021). https://doi.org/10.2478/s11756-020-00501-2

[30] Istock, C. A, N. Ferguson, N. L. Istock and K. E. Duncan. 2001.Geographical diversity of genomic lineages in Bacillus subtilis (Ehrenberg) Cohn sensulato.Org.Divers. Evol., 1:179 191.https://www.gfbshome.de/fileadmin/user_upload/ode2mods/ode/ode01/ode01_000179/article.pdf

[31] Jaccard. P. 1980.Nouvelles recherchessur la distribution florale. Bull. Soc. Vand. Sci. Nat. 44, 223-270. https://www.researchgate.net/publication/242013831

[32] Elamawi, R. M. A. and R. A. S. El Shafey. 2013. Inhibition effects of silver nanoparticles against rice blast disease caused by Magnaporthegrisea. Egypt. J. Agric. Res. 91: 1271-1283. http://www.arc.sci.eg/ejar/UploadFiles/Publications/115476

[33] El-Argawy. E, M.M.H. Rahhal, A. El-Korany, E.M. Elshabrawy and R.M. Eltahan. 2017. Efficacy of Some Nanoparticles to Control Damping-off and Root Rot of Sugar Beet in El-Behiera Governorate. Asian Journal of Plant Pathology, 11 (1): 35-47, 2017. https://scialert.net/abstract/?doi=ajppaj.2017.35.47

[34] Fraternale, D., L. Giamperi and D. Ricci.2003. Chemical composition and antifungal activity of essential oil obtained from in vitro plants of Thymus mastichina L. J Ess. Oil Res.15:278-281. https://www.tandfonline.com/doi/abs/10.1080/10412905.2003.9712142

[35] Al-Taisan, W. A., A. H. Bahkali, A. M. El-gorban and M. A. El-Metwally (2014). Effective influence of essential oils and microelements against Sclerotiniasclerotiorum. Inter. J. Pharmacol. 10: 275-281. https://scialert.net/abstract/?doi=ijp.2014.275.281

[36] SAS (statistical analysis systems). 2014. Version 9.4 of the SAS system for windows. Cary. NC, USA: SAS Institute Inc. http://support.sas.com/documentation/installcenter/en/ikfdtnwinsr/67228/PDF/default/sreq.pdf

[37] Mayer, A.M., E. Harel, R.B. Shaul. 1965. Assay of catechol oxidase: a critical comparison of methods. Phytochemistry 5: 783-789. https://citationsy.com/archives/q?doi=10.1016/S0031-9422(00)83660-2

[38] Miller, G. L. (1959). Use of dinitrosalicylic acid reagent for determination of reducing sugar. Analytical Chemistry, 31(3), 426- 428. http://dx.doi.org/10.1021/ac60147a030

[39] Nassar,A. M. K and I. A. A. Adss.2016. 2,4Dichlorophenoxy acetic acid, abscisic acid, and hydrogen peroxide-induced resistance-related components against potato early blight (Alternariasolani, Sorauer). Ann AgricSci 61(1):15–23. https://www.sciencedirect.com/science/article/pii/S0570178316300070

[40] Dickerson. D.P, S.F. Pascholati, A.E. Hagerman, L.G. Butler and R.L. Nicholson. 1984. Phenylalanine ammonia-lyase and hydroxycinnamate: CoA ligase in maize mesocotyls inoculated with Helminthosporiummaydis or Helminthosporiumcarbonum. Physiological Plant Pathology 25: 111-123. https://www.sciencedirect.com/science/article/abs/pii/004840598490050X

[41] Livak K.J, Schmittgen T.D. 2001. Analysis of relative gene expression data using real-time quantitative PCR and the 2 (−Delta Delta C (T) method. Methods 25:402–408. https://pubmed.ncbi.nlm.nih.gov/11846609/

[42] Stoleru, V.; Inculet, S.-C.; Mihalache, G.; Cojocaru, A.; Teliban, G.-C.; Caruso, G. Yield and Nutritional Response of Greenhouse Grown Tomato Cultivars to Sustainable Fertilization and Irrigation Management. Plants2020, 9, 1053. https://doi.org/10.3390/plants9081053

[43] McGovern, R.J. (2015). Management of tomato diseases caused by Fusariumoxysporum. Crop Protection. 73: 78–92.

[44] Abou-Zeid.N.M, N. A. Mahmoud and R. A. Saleh. 2016. Effect of some Biotic and Abiotic Applications on Control of Fusarium Wilt of Pepper Plants. Egypt. J. Phytopathol., 44(2): 103-118.https://www.researchgate.net/publication/327551316_

[45] Troncoso-Rojas.R, M. E. Báez-Flores, B. Pryor, H. S. García and M.Tiznado-Hernández. 2013. Inter simple sequence repeat polymorphism in Alternaria genomic DNA exposed to lethal concentrations of isothiocyanates. African J. Microbiol. Res. 7(10):838-852. https://academicjournals.org/journal/AJMR/article-abstract/1B06FA420502

[46] Ahmed, I.S.A. and Lee, Y.S. (2015). Nanoparticles as alternative pesticides: Concept, manufacturing and activities. The Korean Journal of Mycology.43(4): 207–215.

[47] Bouwmeester, H.; Dekkers, S.; Noordam, M.Y. (2009). Review of health safety aspects of nanotechnologies in food production. Regulatory Toxicology and Pharmacology. 53: 52–62.

[48] Sekhon, B.S. (2014). Nanotechnology in agrifood production: an overview. Nanotechnology, Science and Applications. 7: 31–53.

[49] Abdel-Aziz, M.M.; Emam, T.M.; Elsherbiny, E.A. Bioactivity of magnesium oxide nanoparticles synthesized from cell filtrate of endobacteriumBurkholderiarinojensis against Fusariumoxysporum. Mater. Sci. Eng. C 2020, 109, 110617.

[50] El-Abeid, S.E.; Ahmed, Y.; Daròs, J.-A.; Mohamed, M.A. Reduced Graphene Oxide Nanosheet-Decorated Copper Oxide Nanoparticles: A Potent Antifungal Nanocomposite against Fusarium Root Rot and Wilt Diseases of Tomato and Pepper Plants. Nanomaterials2020, 10, 1001. https://doi.org/10.3390/nano10051001

[51] Servin A., Elmer W., Mukherjee A., De la Torre-Roche R., Hamdi H., White J. C. (2015). A review of the use of engineered nanomaterials to suppress plant disease and enhance crop yield. J. Nanopart. Res. 17 92 10.1007/s11051-015-2907

[52] Liu, B. K., Xue, Y. F., Zhang, J. T., Han, B., Zhang, J., Suo, X. Y., et al. (2017). Visible-light-driven TiO2/Ag3PO4 teterostructures with enhanced antifungal activity against agricultural pathogenic fungi Fusariumgraminearum and mechanism insight. Environ. Sci. Nano 4, 255–264. doi: 10.1039/c6en00415f

[53] Ahmed I. S. Ahmed, Dil Raj Yadav, and Youn Su Lee (2016). Applications of Nickel Nanoparticles for Control of Fusarium Wilt on Lettuce and Tomato. IJIRSET, 5(5): 7378- 7385. DOI:10.15680/IJIRSET.2016.0505132

[54] Imada K, Sakai S, Kajihara H, Tanaka S, Ito S, 2016. Magnesium oxide nanoparticles induce systemic resistance in tomato against bacterial wilt disease. Plant Pathology 65, 551– 60.

[55] Lian, J., L. Zhao, J. Wu, H. Xiong, Y. Bao, A. Zeb, J. Tang, W. Liu Foliar spray of TiO2 nanoparticles prevails over root application in reducing Cd accumulation and mitigating Cd-induced phytotoxicity in maize (Zea mays L.) Chemosphere, 239 (2020), pp. 1-12

[56] Gohari, G., Mohammadi, A., Akbari, A. et al. Titanium dioxide nanoparticles (TiO2 NPs) promote growth and ameliorate salinity stress effects on essential oil profile and biochemical attributes of Dracocephalummoldavica. Sci Rep10, 912 (2020). https://doi.org/10.1038/s41598-020-57794-1

[57] Elmer.W, R. Torre-Roche, L. Pagano, S. Majumdar, N. Zuverza- Mena, C.Dimkpa, J. Gardea-Torresdey, and J.C. White. 2018. Effect of Metalloid and Metal Oxide Nanoparticles on Fusarium Wilt of Watermelon. Plant Disease 102(7). https://pubmed.ncbi.nlm.nih.gov/30673561/

[58] Siddaiah. C. N, K. V. H. Prasanth, N. R. Satyanarayana, V. Mudili, V. K. Gupta, N. K. Kalagatur, T. Satyavati, X. Dai, J.Chen, A. Mocan, B. P. Singh and R. K. Srivastava. 2018. Chitosan nanoparticles having higher degree of acetylation induce resistance against pearl millet downy mildew through nitric oxide generation. Scientific reports, 8:2485. https://pubmed.ncbi.nlm.nih.gov/29410438/

Made in United States
Orlando, FL
04 November 2026

90206129R00022